Beautiful Birds

PENGUINS

By Bray Jacobson

Gareth Stevens
PUBLISHING

Please visit our website, www.garethstevens.com. For a free color catalog of all our high-quality books, call toll free 1-800-542-2595 or fax 1-877-542-2596.

Library of Congress Cataloging-in-Publication Data

Names: Jacobson, Bray, author.
Title: Penguins / Bray Jacobson.
Description: Buffalo, New York : Gareth Stevens Publishing, 2024. | Series: Beautiful birds | Includes index. | Audience: Grades K-1
Identifiers: LCCN 2023023727 (print) | LCCN 2023023728 (ebook) | ISBN 9781538292563 (library binding) | ISBN 9781538292556 (paperback) | ISBN 9781538292570 (ebook)
Subjects: LCSH: Penguins–Juvenile literature.
Classification: LCC QL696.S473 J325 2024 (print) | LCC QL696.S473 (ebook) | DDC 598.47–dc23/eng/20230526
LC record available at https://lccn.loc.gov/2023023727
LC ebook record available at https://lccn.loc.gov/2023023728

First Edition

Published in 2024 by
Gareth Stevens Publishing
2544 Clinton St.
Buffalo, NY 14224

Editor: Kristen Nelson
Designer: Andrea Davison-Bartolotta

Photo credits: Cover Giedriius/Shutterstock.com; p. 1 T. Terziev/Shutterstock.com; p. 5 axily/Shutterstock.com; pp. 7, 24 (flipper) TETSU Snowdrop/Shutterstock.com; pp. 9, 24 (claw) boyhey/Shutterstock.com; p. 11 Sergey Uryadnikov/Shutterstock.com; p. 13 Goinyk Production/Shutterstock.com; p. 15 slowmotiongli/Shutterstock.com; p. 17 Rich Lindie/Shutterstock.com; p. 19 Phil West/Shutterstock.com; pp. 21, 24 (chick) vladsilver/Shutterstock.com; p. 23 MagicBloods/Shutterstock.com.

Printed in the United States of America

CPSIA compliance information: Batch #CW24GS: For further information contact Gareth Stevens, New York, New York at 1-800-542-2595.

Contents

Penguins are birds.
They live near water.

They do not fly.
They swim well.
They use flippers.

They have webbed feet.
Feet have claws.

They have feathers.
These keep skin dry.

They have black backs.
They have white bellies.

They hunt for fish.
They eat shrimp.

Some kinds make nests.
Chinstraps use stones.

Mothers lay eggs.
Eggs are white, blue,
or green.

Babies are
called chicks.
They look different
from adults.

Most chicks
have feathers.
They may be white
or black.
They may be brown
or gray.

Words to Know

chick

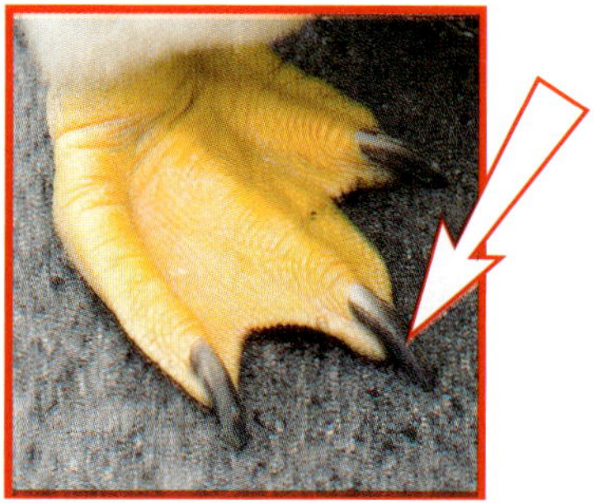

claw

flipper

Index